Everyday Mathematics®

The University of Chicago School Mathematics Project

My First Math Book

Kindergarten

 **Wright Group**

The **McGraw·Hill** Companies

The University of Chicago School Mathematics Project (UCSMP)
Max Bell, Director, UCSMP Elementary Materials Component; Director, *Everyday Mathematics* First Edition
James McBride, Director, *Everyday Mathematics* Second Edition
Andy Isaacs, Director, *Everyday Mathematics* Third Edition
Amy Dillard, Associate Director, *Everyday Mathematics* Third Edition

Third Edition Early Childhood Team Leaders
David W. Beer, Deborah Arron Leslie

Technical Art
Diana Barrie

Teachers in Residence
Ann E. Audrain, Dorothy Freedman, Margaret Krulee, Barbara Smart

Editorial Assistant
Patrick Carroll

Contributor
John Saller

www.WrightGroup.com

 **Wright Group**

Printed in the United States of America.

Send all inquiries to:
Wright Group/McGraw-Hill
P.O. Box 812960
Chicago, IL 60681

ISBN 0-07-604524-2

5 6 7 8 9 QW-E 12 11 10 09 08 07

The *McGraw·Hill* Companies

Contents

Contents

Contents

Contents

Contents

Craft-Stick Patterns

1. Draw your craft-stick pattern here.

2. Draw another craft-stick pattern.

Measuring with My Foot

1. Measure a table with your cutout foot.

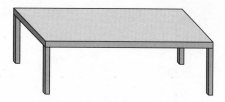

Record your measurement.

_____ my feet

2. Measure and draw another object.

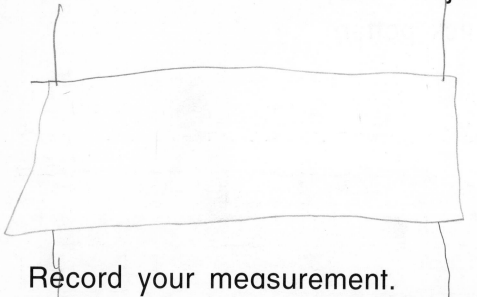

Record your measurement.

_____ my feet

2

Dice-Throw Grid

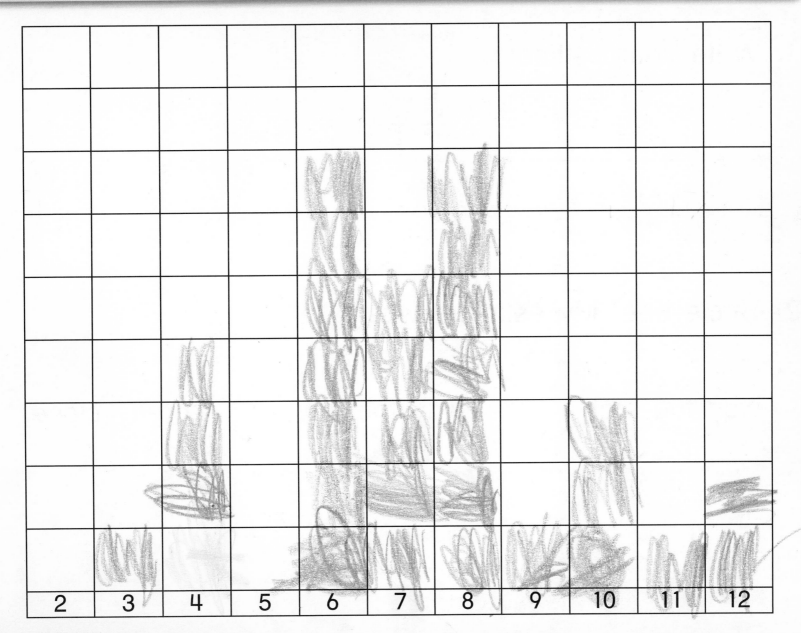

| 2 | 3 | 4 | 5 | 6 | 7 | 8 | 9 | 10 | 11 | 12 |

Use with Activity 5•8.

3

1. Write your estimate.

_____

2. Circle one. My estimate was

way too high.   way too low.   pretty close.

4

Comparing Measurements

1. Draw the object you and your partner are measuring.

2. Measure and draw another object in the room.

Record your measurements.

_____ my feet

_____ my partner's feet

_____ standard feet

_____ my feet

_____ my partner's feet

_____ standard feet

Measuring with Standard Measuring Tools

1. Use a standard measuring tool to measure an object.

   Draw the object you measured.

2. Record your measurement. _____

3. Draw the measuring tool you used.

6

Pet Bar Graph

1. Draw one kind of pet from the graph.

2. How many children have this pet? _____

3. Show one more thing you learned from the graph.

Penny Piggy Bank

Draw your pennies in the piggy bank. Write how many pennies you have.

Using Counting as a Measure of Time

1. How many counts did it take you to cross the room? Write the number below.

    walking _____ counts

    tiptoeing _____ counts

2. Choose a new way to cross the room. _____

3. Write the number of counts it took. _____ counts

4. Which way took the longest time? _____

5. Which way took the shortest time? _____

Measuring with Different Tools

1. Measure an object using different tools.
Draw the object you measured.

2. Record measurements for each tool you used.
Write the measurement unit, such as cubes or my feet.

_____ _____
units

_____ _____
units

10

Recording Estimates

1. Write your estimate.

_____

2. Circle one. My estimate was

way too high.   way too low.   pretty close.

Recording Half Groups

Divide a handful of counters into 2 equal groups.
Draw counters in the boxes to show the 2 equal groups.

Use with Activity 6·11.

Showing Patterns with Symbols

1. Think of a movement pattern.

Draw symbols to show your pattern.

2. Show your pattern to someone.

Tell what the symbols mean.

Have him or her follow the pattern.

Tallying Coin Flips

Make a tally mark each time you get "heads."

Make a tally mark each time you get "tails."

Class Collection

1. Our class is collecting _____.

2. Record how the class collection is growing.

Date	How many **new** items did we add?	How many items do we have **all together**?

Dice-Throw Grid

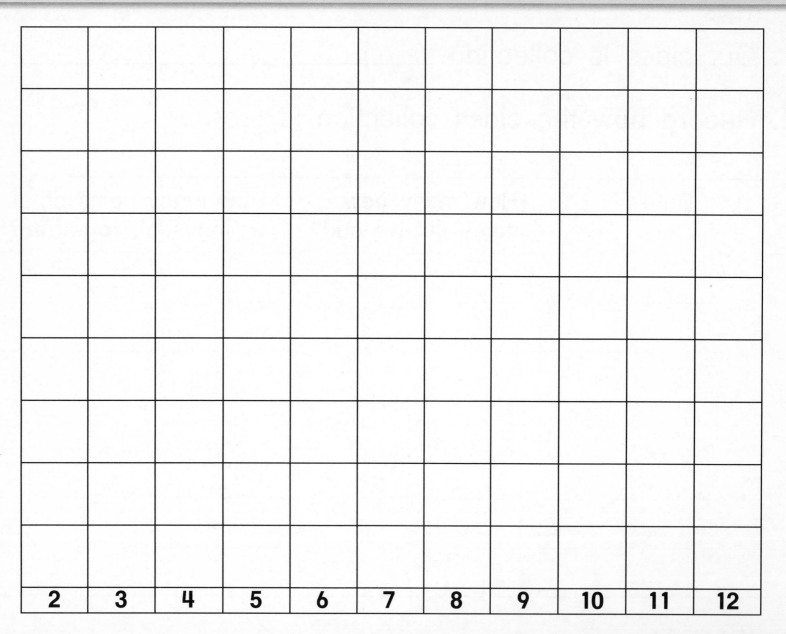

| 2 | 3 | 4 | 5 | 6 | 7 | 8 | 9 | 10 | 11 | 12 |

16

Number Story

Think of a number story.

Show your story with pictures and a number sentence.

10s and 1s with Craft Sticks

Bag 1

1. How many bundles of 10 do you have? _____

2. How many unbundled sticks do you have? _____

3. How many sticks do you have all together? _____

Bag 2

1. How many bundles of 10 do you have? _____

2. How many unbundled sticks do you have? _____

3. How many sticks do you have all together? _____

18

Recording Estimates

1. Write your estimate.

_____

2. Circle one. My estimate was

way too high.　　　way too low.　　　pretty close.

Double-Digits with Dice

1. Write the two digits you rolled.

_____ and _____

2. Write both 2-digit numbers you can make with

those digits.

_____ and _____

3. Circle the larger number.

20

Bead String Name Collections

1. Write the number of beads on your counting loop.

_____

2. Draw 3 different ways you grouped the beads on your loop.

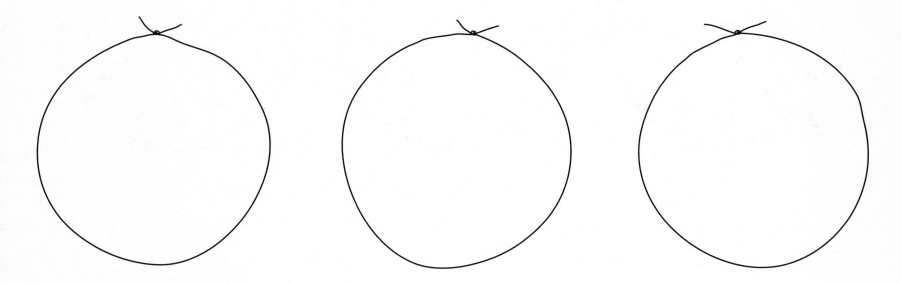

Hour-Hand Clock

Draw the hour hand on each clock to match the time your teacher says.

1.

2.

3.

Use with Activity 8·3.

Weather Graph Questions

Use the class **weather graphs** to answer these questions.

1. What month had the most sunny days? _____

2. What month had the least sunny days? _____

3. Did any months have 0 rainy days? If so, which ones?

_____   _____   _____

4. Write a question about the weather graphs for someone else to answer.

Function Machines

1. Follow the rule and write the *out* numbers.

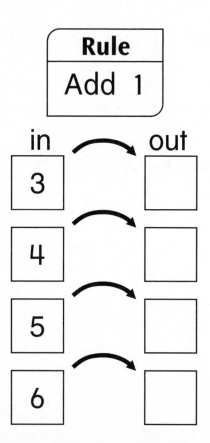

Rule
Add 1

in → out

in	out
3	
4	
5	
6	

2. Make your own rule. Write the *in* numbers. Have a partner follow your rule and write the *out* numbers.

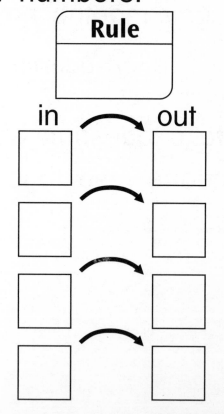

Rule

in → out

in	out

24

Use with Activity 8·7.

1. Write the missing number.

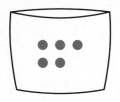 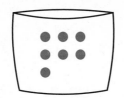

5 + ___ = 7

2. Write the missing number.

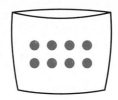 

8 − ___ = 4

3. Make up an addition pocket problem.

___ + ___ = ___

4. Make up a subtraction pocket problem.

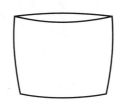 

___ − ___ = ___

Draw what you put in both sides of your pan balance to make it **level.**

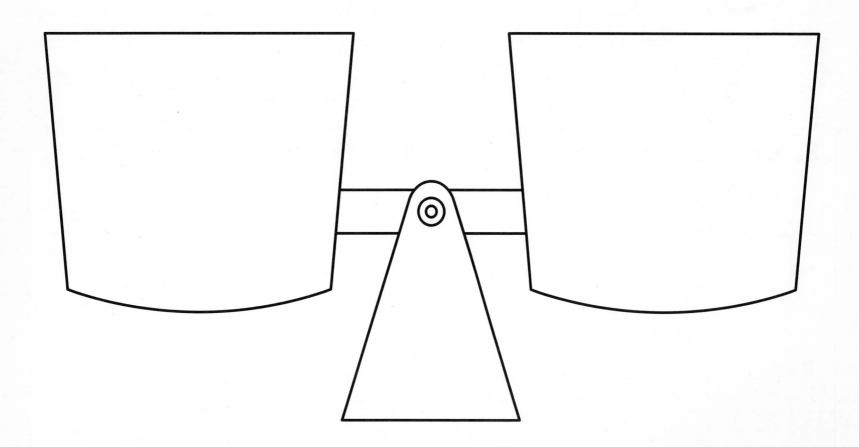

Use with Activity 8•15.

Number Writing (0)

Name _____

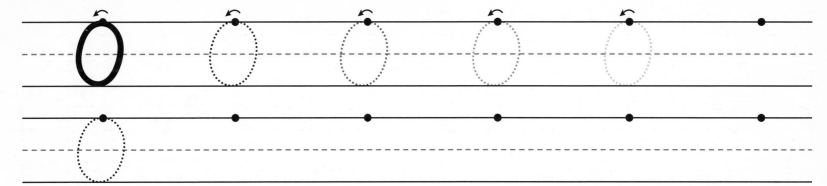

Number Writing (1)

Name _____

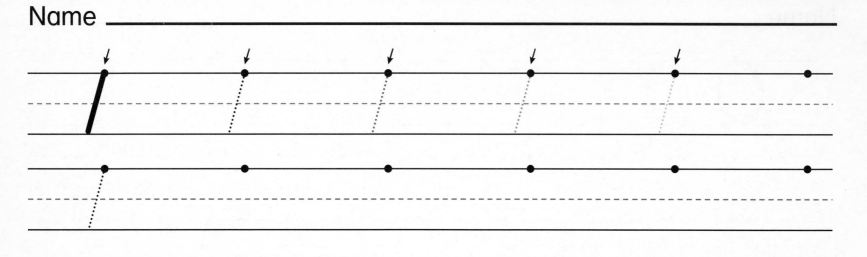

Use with Activity 8•15.

Number Writing (2)

Name _____

Name _____

3   3   3   3   3   •

3   •   •   •   •   •

Number Writing (4)

Name _____

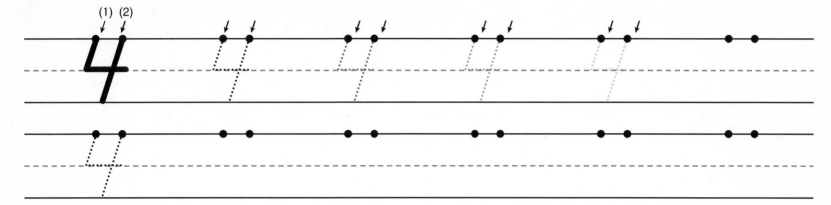

Name _____

5   5   5   5   5

5

Number Writing (6)

Name _____

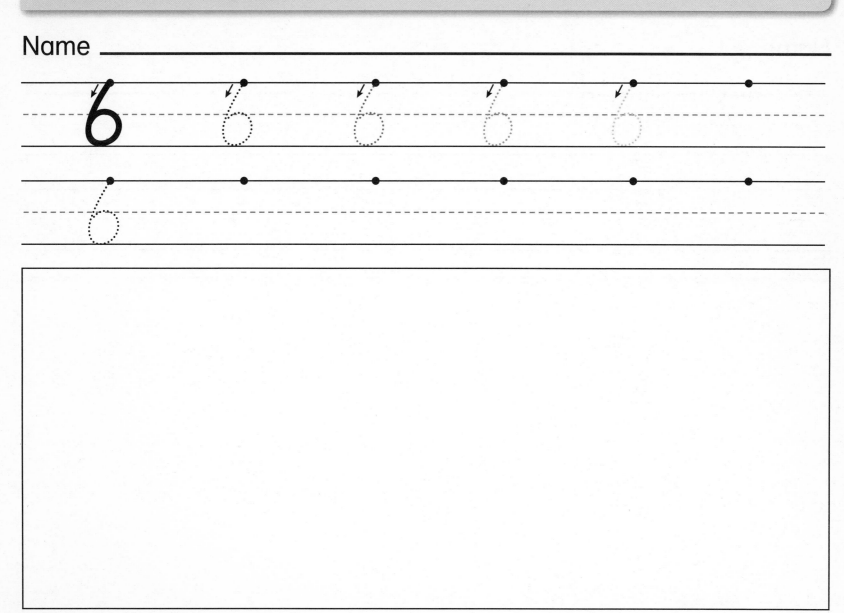

Number Writing (7)

Name _____

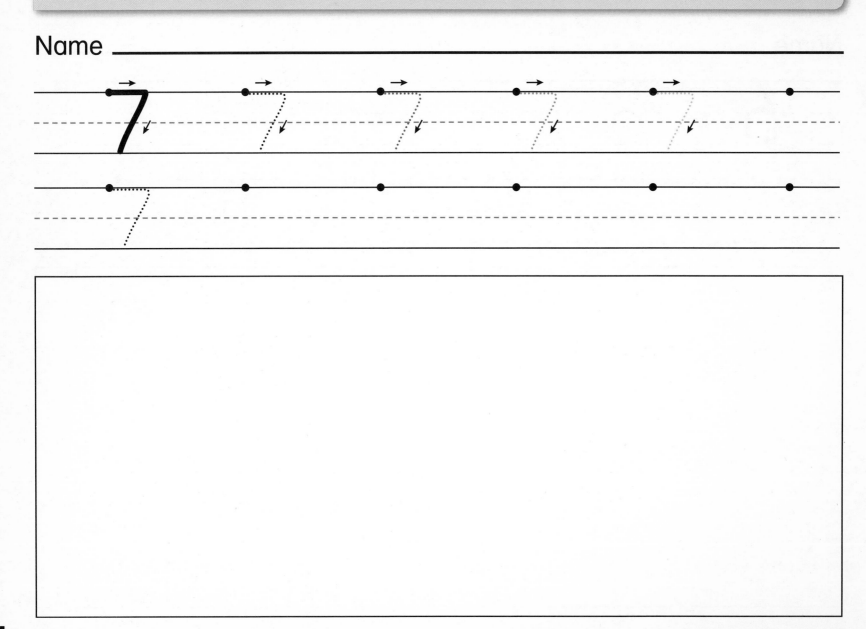

Number Writing (8)

Name _____

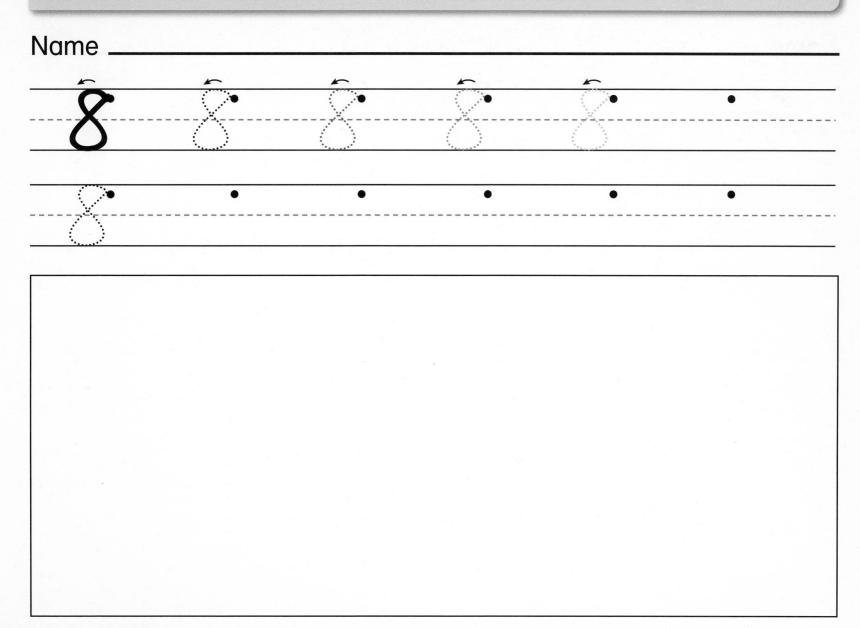

Name _____

q   q   q   q   q   •

q   •   •   •   •   •

Activity Sheet 4

Number Writing (10)

Name _____

10   10   10   10   10

10

Name _____

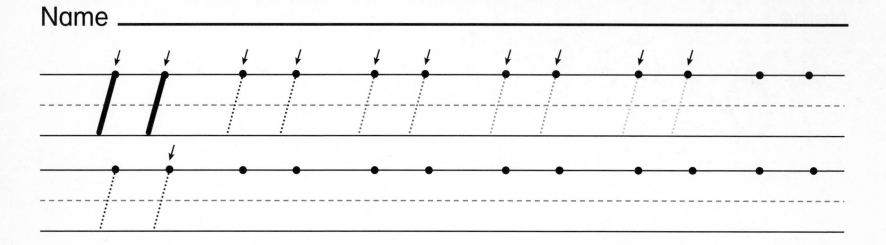

Use with Activity 8·15.

Number Writing (12)

Name _____

Number Writing (13)

Name _____

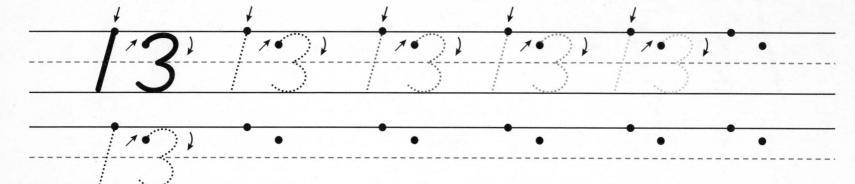

Number Writing (14)

Name _____

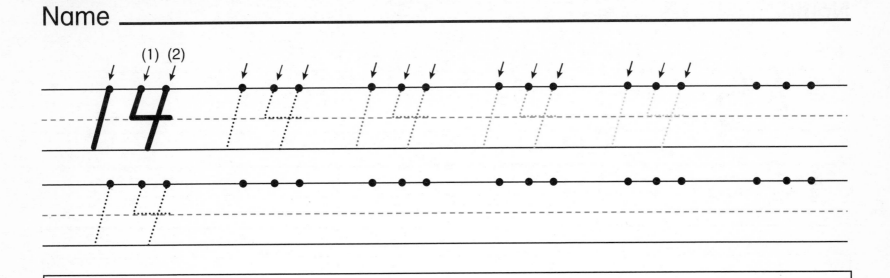

Number Writing (15)

Name _____

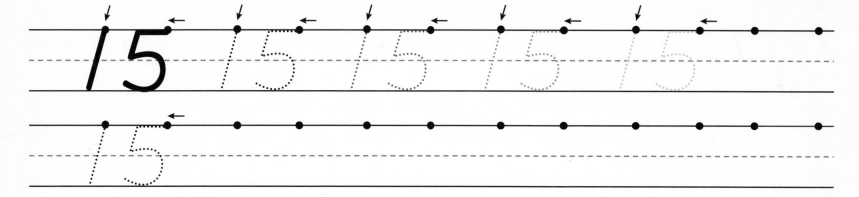

42

Number Writing (16)

Name _____

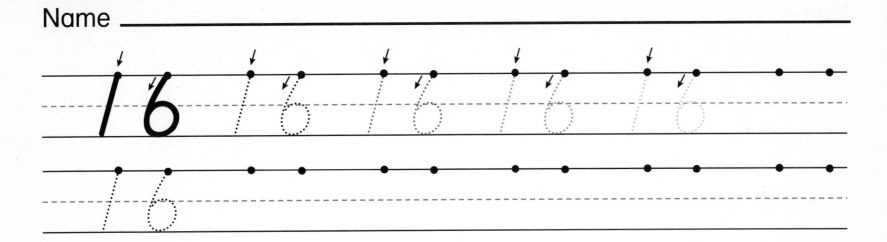

Use with Activity 8 • 15.

Name _____

Use with Activity 8•15.

Name _____

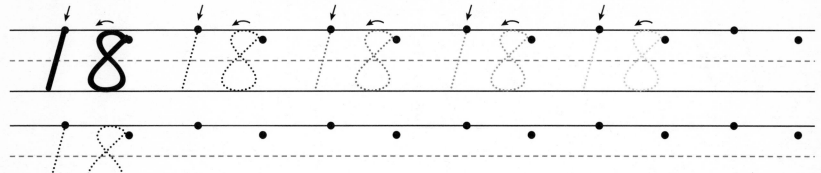

Number Writing (19)

Name _____

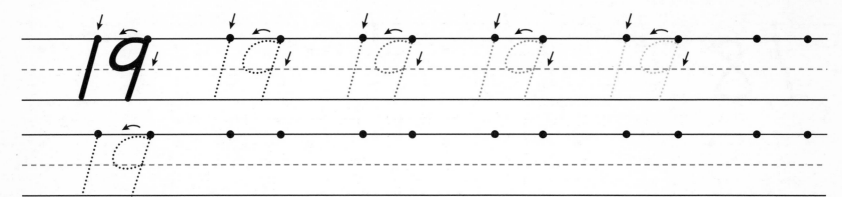

Number Writing (20)

Name _____

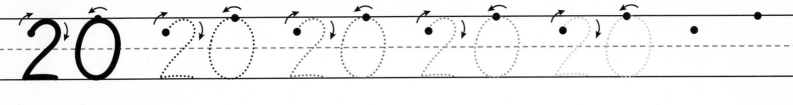

Use with Activity 8 • 15.

Writing and Drawing Page

Used with Activity _____

Writing and Drawing Page

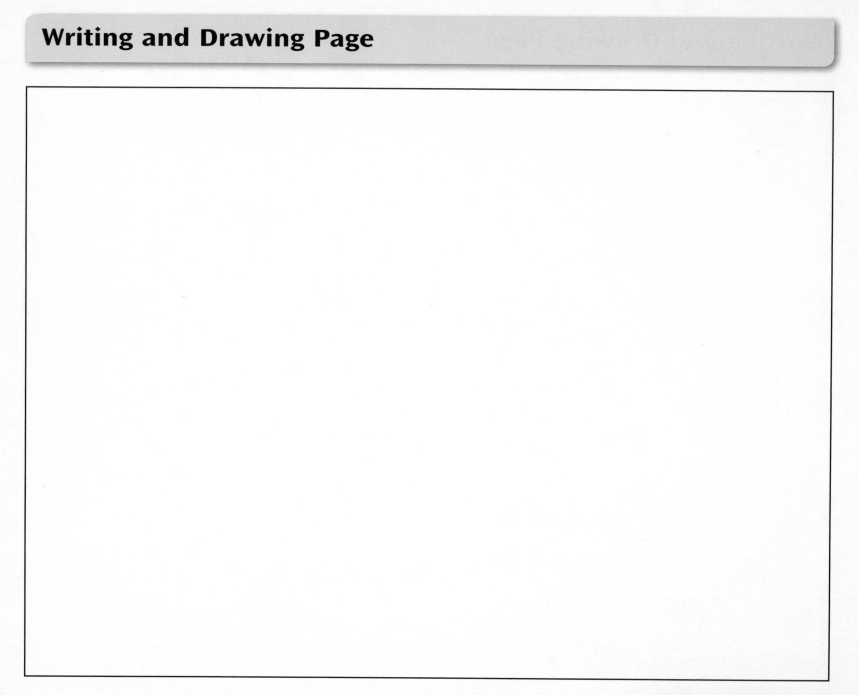

Used with Activity _____

Writing and Drawing Page

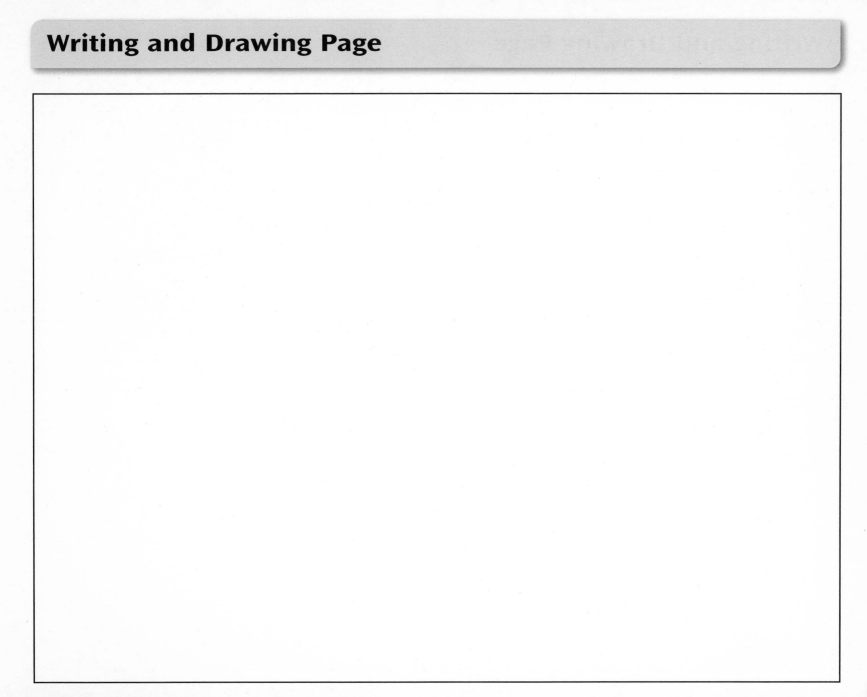

Used with Activity _____

Writing and Drawing Page

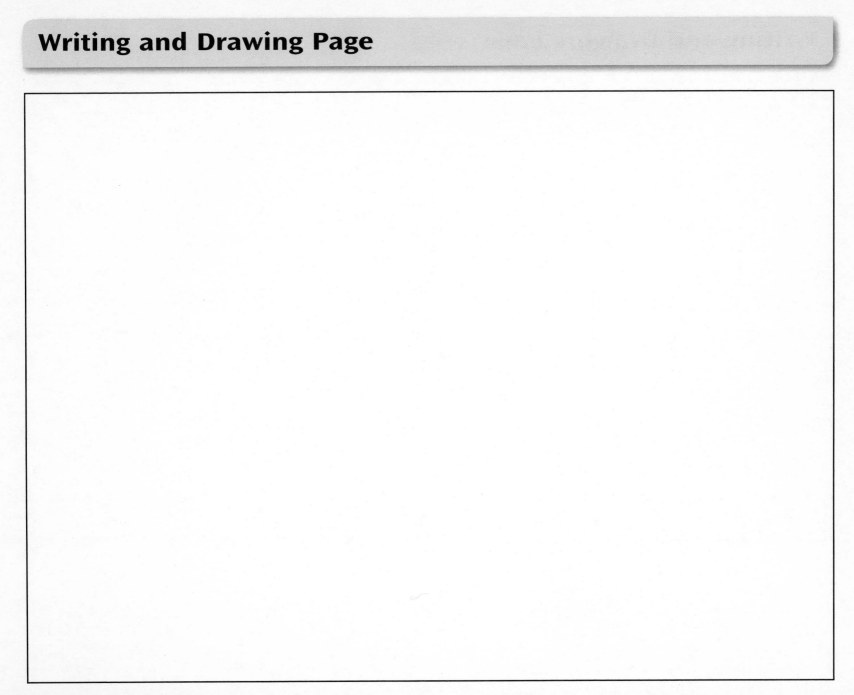

Writing and Drawing Page

Used with Activity ____

Writing and Drawing Page

Writing and Drawing Page

Writing and Drawing Page

Writing and Drawing Page

Used with Activity _____

Writing and Drawing Page

Writing and Drawing Page

Writing and Drawing Page

Used with Activity _____

Writing and Drawing Page

Writing and Drawing Page

Used with Activity _____

Writing and Drawing Page

Used with Activity _____

Writing and Drawing Page

Writing and Drawing Page

Writing and Drawing Page

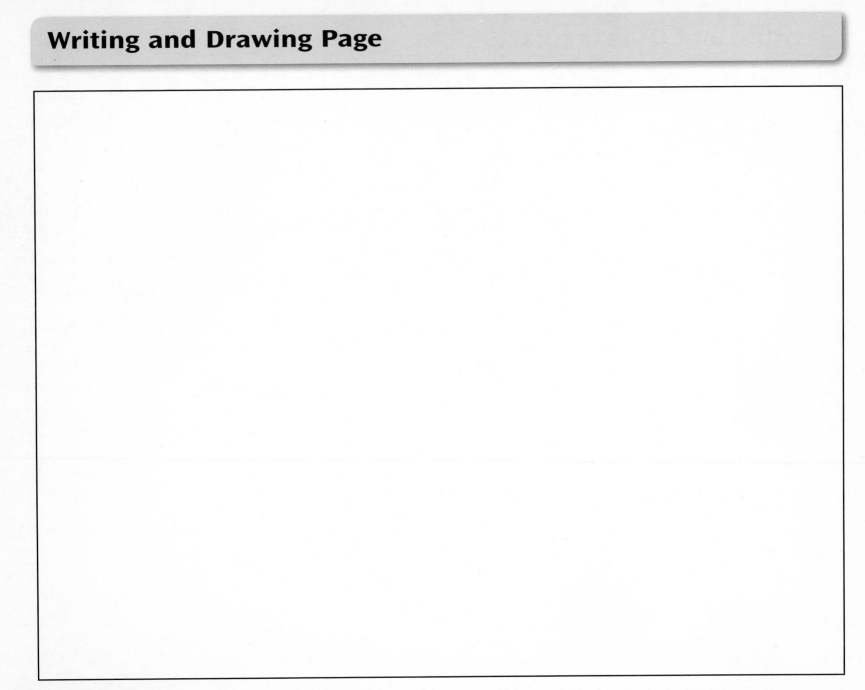

Writing and Drawing Page

Writing and Drawing Page

Writing and Drawing Page

Writing and Drawing Page

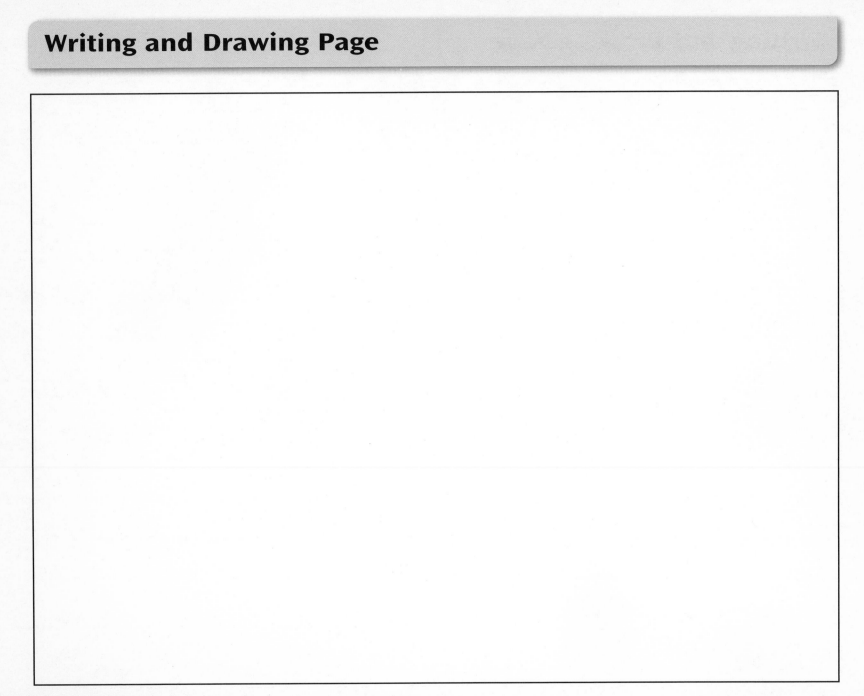

Writing and Drawing Page

Writing and Drawing Page

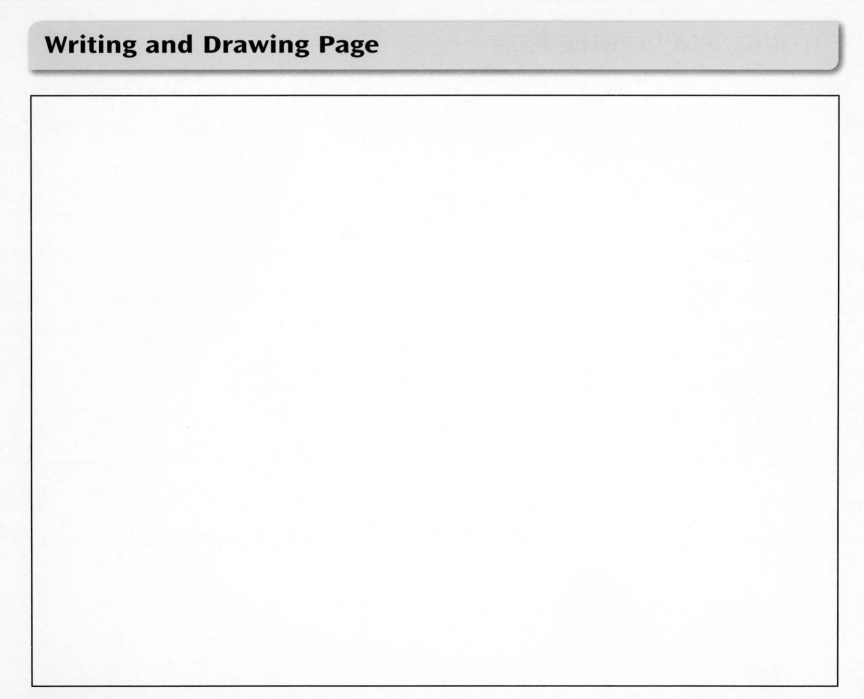

Writing and Drawing Page

Used with Activity _____

Writing and Drawing Page

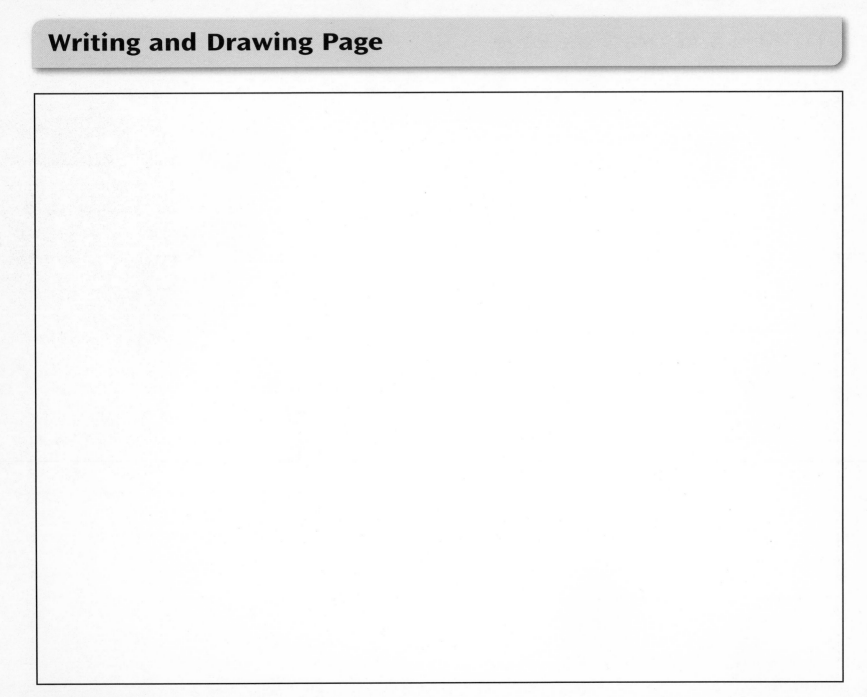

Writing and Drawing Page

Used with Activity _____

Number Grid

									0
1	2	3	4	5	6	7	8	9	10
11	12	13	14	15	16	17	18	19	20
21	22	23	24	25	26	27	28	29	30
31	32	33	34	35	36	37	38	39	40
41	42	43	44	45	46	47	48	49	50
51	52	53	54	55	56	57	58	59	60
61	62	63	64	65	66	67	68	69	70
71	72	73	74	75	76	77	78	79	80
81	82	83	84	85	86	87	88	89	90
91	92	93	94	95	96	97	98	99	100
101	102	103	104	105	106	107	108	109	110

2-Dimensional Shapes

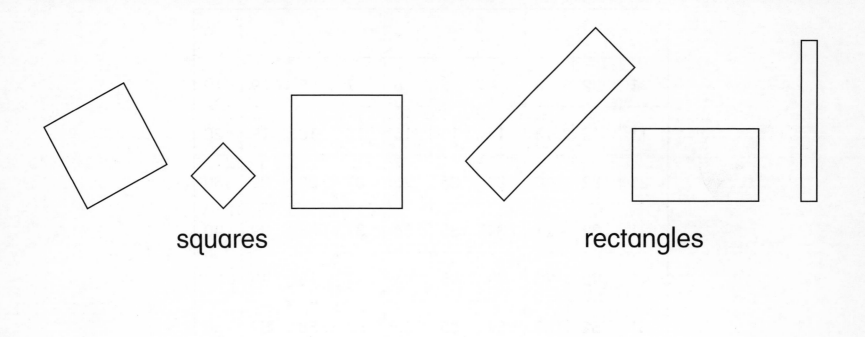

squares

rectangles

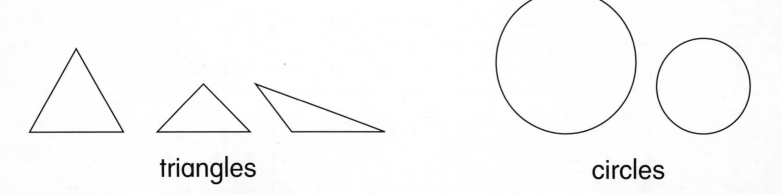

triangles

circles

3-Dimensional Shapes

sphere

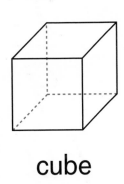

cube

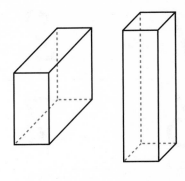

rectangular prisms

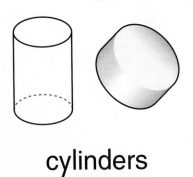

cylinders

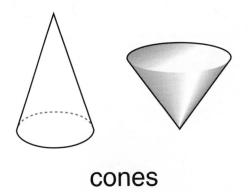

cones

1	2	3	4	5
6	7	8	9	10
11	12	13	14	15
16	17	18	19	20